THE WEATHER REPORT

Seasons and Patterns

JOHN HOPKINS

PERFECTION LEARNING®

Editorial Director: Susan C. Thies
Editor: Mary L. Bush
Design Director: Randy Messer
Book Design: Michelle Glass, Lori Gould
Cover Design: Michael A. Aspengren

A special thanks to the following for their scientific review of the book:

Judy Beck, Ph.D.; Associate Professor, Science Education; University of South Carolina-Spartanburg

Jeffrey Bush; Field Engineer; Vessco, Inc.

IMAGE CREDITS:
©Reuters NewMedia Inc./CORBIS: p. 5 (bottom right); ©Skylab/NRL/NASA/Roger Ressmeyer/CORBIS: p. 8 (bottom left); ©Jamie Harron; Papilio/CORBIS: p. 12 (bottom); ©Layne Kennedy/CORBIS: p. 17 (bottom left); ©Kelly-Mooney Photography/CORBIS: p. 19 (bottom right); ©Tom Stewart/CORBIS: p. 20 (top left); ©AFP/CORBIS: p. 23 (top right); ©FK Photo/CORBIS: p. 25 (bottom right); ©Chase Swift/CORBIS: p. 26 (top); ©Bob Winsett/CORBIS: p. 30 (bottom left)

Corel: cover, pp. 1, 4, 5 (background), 8 (background), 11, 12 (background), 14 (top), 19 (background), 20 (top middle and right), 21, 25 (background), 27 (top and middle), 28, 29, 30 (background), 31, 33, 35 (right), 38–39, 40, all art behind sidebars; PhotoDisc: pp. 2–3, 27 (bottom), 34 (left), 37; Digital Stock: pp. 15, 16 (top), 17 (top), 24, 34 (right), 35 (left); Perfection Learning Corporation: pp. 6, 7, 9, 10, 14 (bottom), 18 (bottom), 22, 32; ©Royalty-Free/CORBIS: p. 36

For information, contact
Perfection Learning® Corporation
1000 North Second Avenue, P.O. Box 500
Logan, Iowa 51546-0500.
Phone: 1-800-831-4190
Fax: 1-800-543-2745
perfectionlearning.com

1 2 3 4 5 6 BA 08 07 06 05 04 03

ISBN 0-7891-6016-1

Contents

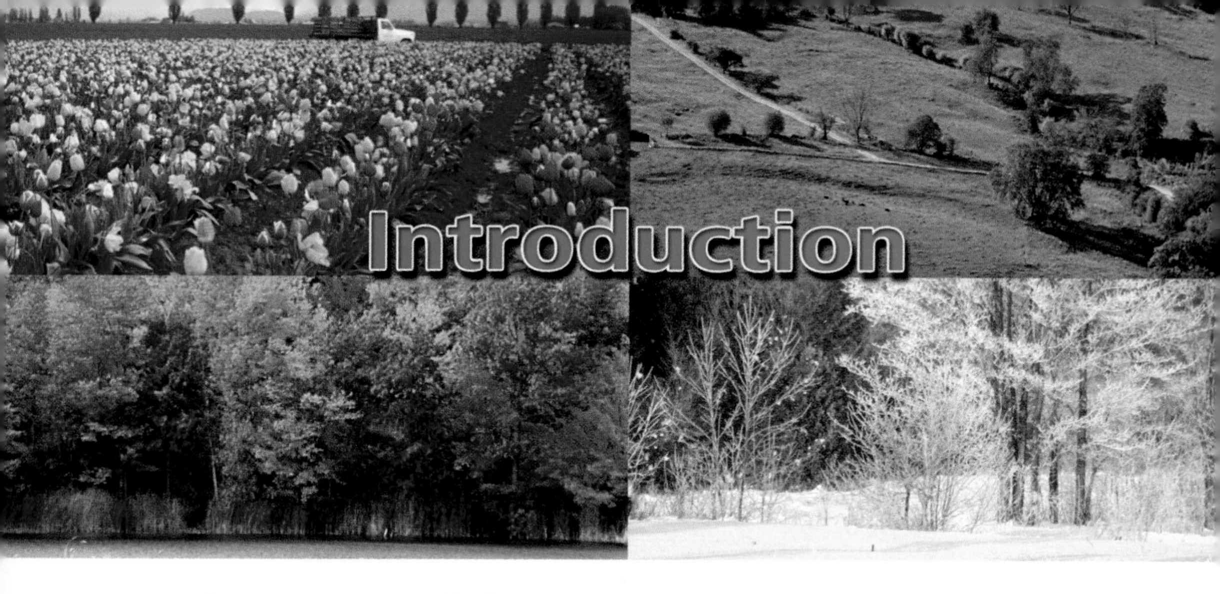

Introduction

What comes to mind when you think of seasons? Snowball fights, sledding, and hot chocolate? Swimming pools, barbecues, fresh fruit, and baseball? How about hayrides and pumpkins? Don't forget spring break. But there is much more to the seasons than food, fun, and pastimes enjoyed with family and friends.

Depending on the region of the world, each season means a change in temperatures and **precipitation**. Usually, a change in the pattern or type of weather also occurs.

In most places, summer is warmer and drier with longer days. Fall, or autumn, brings cooler, shorter days. Winter means very short, cold days. The days begin to lengthen in the spring, bringing blooming flowers and blossoming trees. Temperatures grow milder. Snowstorms give way to rainstorms. The world thaws and awakens. Summer is nearing and we've come full circle.

Every year, the seasons and patterns repeat themselves. Each area of the world comes to expect a certain kind of weather during its seasons. Some years will bring more **extreme** weather. There will be heat waves and drought or roaring blizzards and floods. But one fact remains. Summer is always followed by fall, fall by winter, winter by spring, and spring by summer. These seasons will continue to guide our celebrations, lifestyles, and **environments**.

1

Mother Earth and Bad Posture

The planet Earth is shaped like a big ball. It is a giant, spinning **sphere**. The Earth completes one full turn, or rotation, every 24 hours. This spinning creates day and night. When the side of the planet you live on faces the Sun, it's daytime. As you spin away from the Sun, it becomes night.

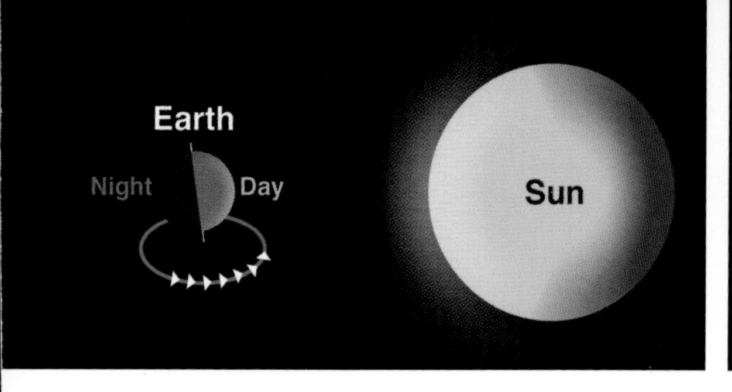

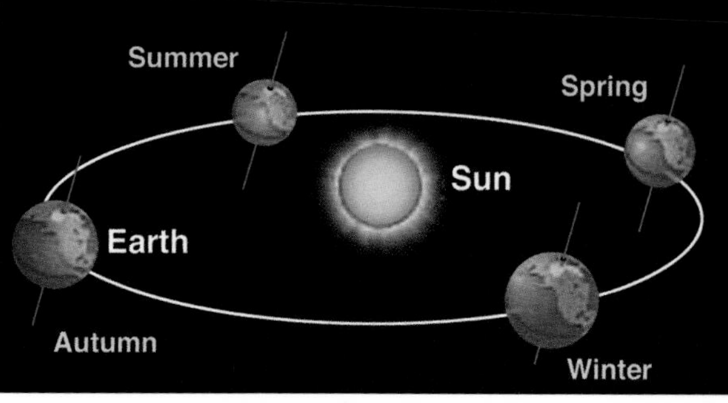

The Earth also **orbits** around the Sun. It is part of a **solar system**. The Sun is near the middle of this solar system. It does not move, but everything else does. Planets, moons, and asteroids travel around the Sun. It takes the Earth one year to complete its journey around the Sun. This is called a *revolution*. The seasons change according to where the Earth is in its orbit.

The Earth's Axis

Have you ever been told to stand up straight and stop leaning? While leaning is bad for your body, a leaning or tilted Earth is important to the changing seasons. Imagine a long stick through the middle of the Earth. The Earth spins on this imaginary stick, which is called an **axis**. The North and South Poles are at the ends of the Earth's axis.

The Earth's axis is slanted. No matter where the Earth is in its orbit, the tilt, or angle, of the axis remains the same. As the Earth orbits, part of it leans toward the Sun and part leans away. The part of the planet leaning toward the Sun enjoys summer. The part leaning away experiences winter. Spring and autumn fall in between.

The Equator

The middle of the Earth is always the same distance from the Sun. This part is called the *equator*. It's almost always warm near the equator, and the seasons don't change much. Move north or south and the farther you go, the greater the change in the seasons.

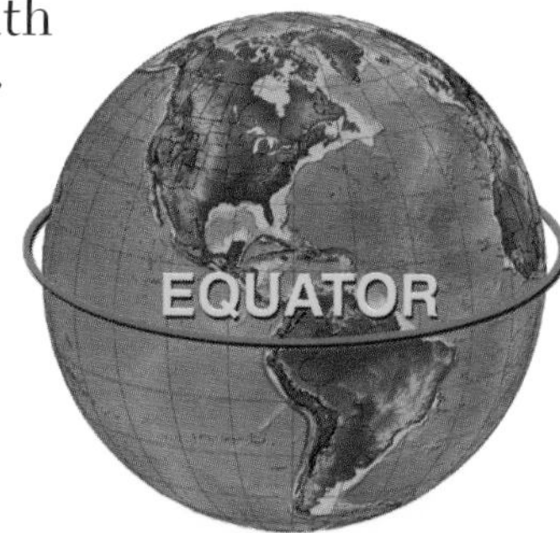

Materials

flat surface
flashlight or lamp
disposable tablecloth
pencil
apple
sharp skewer or stick long enough to push through the apple

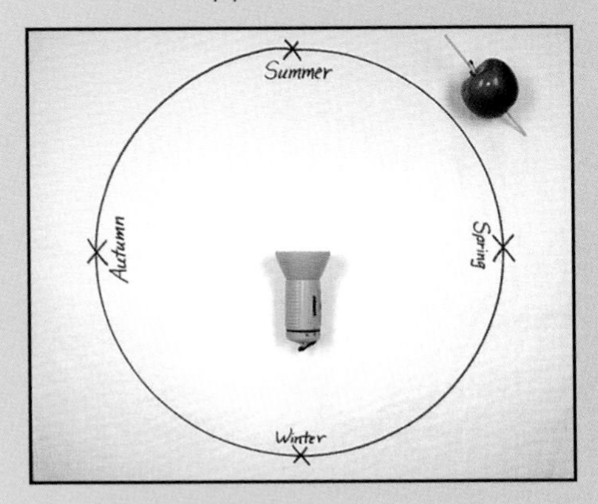

Procedure

Draw a circle on the tablecloth. Make four Xs on the circle, each one a quarter of the way around. Label one X *summer*. Moving counter-clockwise, label the next X *autumn*, followed by *winter* and *spring*. Make sure that winter and summer are on opposite sides of each other.

Carefully push the stick through the center (core) of the apple. One end should come out where the apple's stem is. The other end should come out in the center of the bottom of the apple.

The apple is the Earth, and the flashlight or lamp is the Sun. Put the light in the middle of the circle. Turn it on and darken the room. Now place the apple in the "summer" position. Point the top of the stick toward the lamp. The top half of the apple is tilted closest to the lamp. When the Earth is tilted in this position, it is summer at the top and winter at the bottom. Can you see that the bottom of the stick is tilted away from the Sun?

Keeping the stick pointed in the same direction, move the apple around the circle, stopping at each X. Where is the top of the stick pointing? The bottom? Pay close attention to where the light (Sun) hits the apple (Earth) at each X. Record your observations on a separate sheet of paper.

**Special note: Save your materials for Chapter 2.

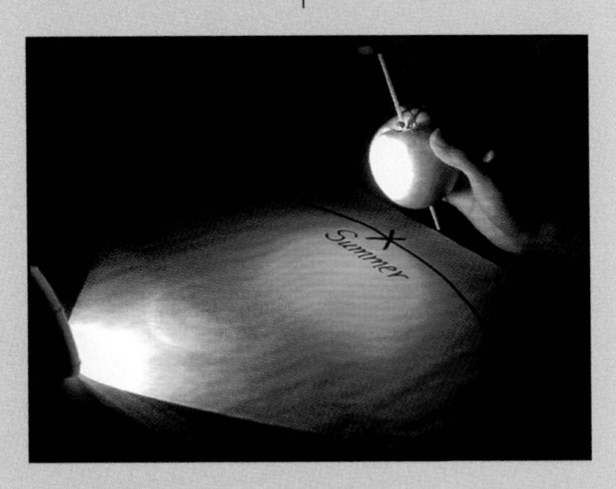

Hemispheres, Climates, and the Equator

Hemispheres

Because the Earth's axis is tilted, either the northern or southern half will almost always be leaning closer to the Sun. When the upper half, or Northern **Hemisphere**, leans closest to the Sun, it is summer there. When the lower half, or Southern Hemisphere, leans closest to the Sun, it is summer in that hemisphere. Whatever season is occurring in the Northern Hemisphere, the opposite season is occurring in the Southern Hemisphere. While you build a snowman in the North, a friend builds a sand castle in the South.

Return to the apple/seasons project from Chapter 1. On your tablecloth, write *winter* below the word *summer*. For each of the remaining three seasons, write the opposite season beneath.

Now move your apple (Earth) around the light (Sun) again in the same tilted position. At each X position, what season is each hemisphere experiencing?

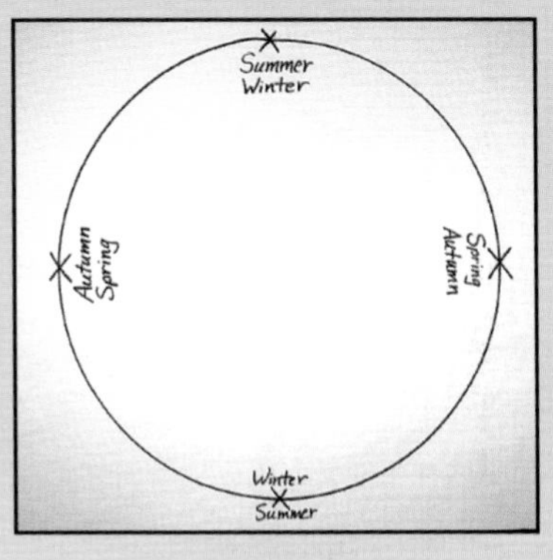

Climates

Did you ever wonder why it's icy cold at the North and South Poles? Or why the rain forests near the equator are steamy? The tilt of the Earth's axis is largely responsible for the **climate** in each area of the world.

Picture the North Pole (the Arctic) at the top of the Earth's axis. The South Pole (the Antarctic) is at the bottom. The Poles never receive direct sunlight, so the temperatures there are always chilly. Instead, the Sun shines its light directly between the Poles—on the equator. This area receives the most heat from the Sun, so its temperatures are warmer.

Since countries along the equator are so hot, you might expect them to be dry. Not so! In many areas near the equator, it rains every day. Water **evaporates** much easier when the air is warm. As moisture evaporates, it doesn't disappear. The water vapor sticks to small pieces of dust, soot, or sea salt in the air and forms droplets. These droplets form clouds and continue to grow. Like a sponge, a cloud can only hold so much water. Eventually the water will return to Earth as rain. If there is a lot of water evaporating in an area, it will rain frequently.

You might also think that with all the ice at the Poles, the **polar** weather would be very wet. In fact, the North and South Poles are fairly dry. Cold air does not pick up and hold water as easily as warm air does.

However, when cold air flowing from the Poles mixes with warm, wet air flowing from the equator, it's time to get out the snow shovels! The warm, moist air condenses into droplets that freeze into clouds. These droplets continue to grow as more moisture freezes to them. When the ice droplets fall back through the warmer air, they melt into a raindrop. But if the air near the Earth's surface is cold, the ice won't melt and it will fall as snow.

Climatic Zones

If you traveled north from the South Pole, you would find the weather warming gradually. Once you passed over the equator, conditions would become cooler the farther north you went. As you move north or south through changing weather patterns, you are moving through **climatic** zones, or regions. These zones are called *tropical*, *temperate*, and *polar*.

Draw a circle with a straight line across the middle. Write the word *equator* on the line. Draw two more lines above the equator and two more below it. Leave the same amount of space between each line and at the top and bottom of the circle.

Starting at the equator, write the word *tropical* in the space (zone) above and below the center line. Next, write the word *temperate* in the spaces above and below *tropical*. Finally, in the spaces at the top and the bottom of the circle, write *polar*.

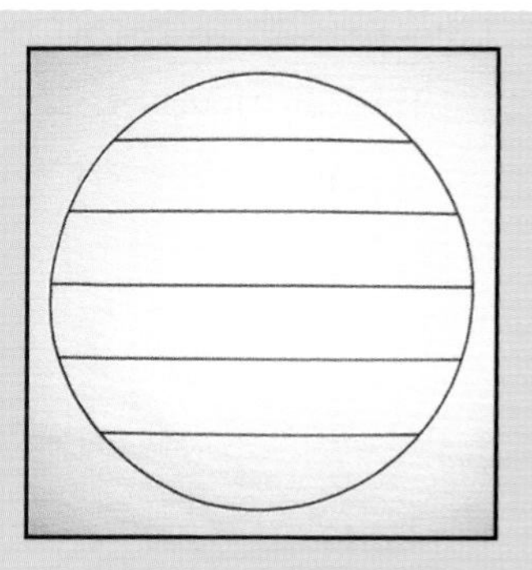

The equator and the tropical zones receive the most sunlight and are the warmest. Polar zones are the coldest. The temperate zones fall in between the two since they have a milder climate. If you don't like it too hot or too cold, the temperate zone is where you want to be.

Locate where you live on a globe. In what climatic zone do you live?

Changes Within a Zone

Traveling east or west across the globe, you stay in the same climatic zone, experiencing similar weather patterns. However, different areas in a zone may vary. Deserts in a zone will have hotter weather, while mountains are cooler. Huge bodies of water and large jungles also affect the weather. Some places along the equator are hot, dry, and **barren**. Others are **humid** rain forests, thriving with life. All of this is true because the land and water features in an area can also affect its weather.

The surface of a region can make an area hotter or colder. Surfaces such as rock, dirt, sand, or concrete **reflect** heat, making the air hotter. Grass and water **absorb** some of the heat, making the air cooler.

Plants can also change an environment. Lots of trees provide shade for an area, which keeps it cooler. Large, tall trees can even completely block out the Sun's rays in some places, making it even colder.

The higher in **elevation** a region is, the thinner and colder the air gets. Because of this, mountain areas are often colder than the surrounding land.

So whether you live in a tropical, temperate, or polar zone, you must also consider the unique environments found within each one.

3

The Big Thaw

Birds chirp and the year's first flowers burst with color. Bees busily gather nectar and spread pollen. Life emerges from cocoons, hives, and burrows to greet the warming season. Snow and ice give way to rivers, streams, lakes, and oceans. It's spring!

The Vernal Equinox

The first day of spring is called the *vernal equinox*. The word *equinox* means "equal night." On the vernal equinox, day and night are the same length. It happens around March 21 in the Northern Hemisphere and around September 22 in the Southern Hemisphere.

Signs of Spring

As spring gets underway, the days begin to lengthen and the air warms. Soon nature responds. Insects buzz and flowers bloom.

The Earth's surface slowly begins to warm. Water warms more slowly than land. The warming atmosphere causes wind patterns to change. Moisture begins to evaporate from the Earth's surface.

Spring Equinox

The word *vernal* comes from the Latin word *vernalis*, which means "spring."

Opposite Equinoxes

When one hemisphere is experiencing the vernal equinox, the other welcomes the autumnal equinox.

Check the weather forecast, and choose a day when sunshine is predicted. In the late afternoon or early evening, place a large pan or small pool on a concrete, asphalt, or flat dirt surface. Make sure that it is in an area that will not be bothered overnight and will get direct sunlight the next day. Fill the pool or pan with water and leave it overnight.

Early the next morning, check the temperature of the water and the temperature of the surface next to the container. Check both again about midday. Which is warmer? Check again later in the afternoon while the area is still receiving sunlight. Which is warmer now? Which is heating up more quickly?

Record your observations, and write a conclusion about what the experiment showed you about warming land and water.

Rainfall

What goes up eventually comes down. When water evaporates, it rises into the **atmosphere**. Wind carries it to various locations where it falls as rain. The amount of water on Earth always stays the same. The only thing that changes is how weather patterns spread it around from season to season and year to year. This is known as the **hydrologic cycle**.

In certain areas, thunderstorms and even tornadoes can result as warmer air picks up evaporating water. Warm air trying to move upward can push cold air away, creating wind or changing the wind's direction. So spring storms often bring pouring rains and heavy winds.

Spring thawing and rainfall help revive the plants and wildlife. Gardens are planted. Flowers poke out of the soil. Animals drink from rivers and streams. Life begins anew!

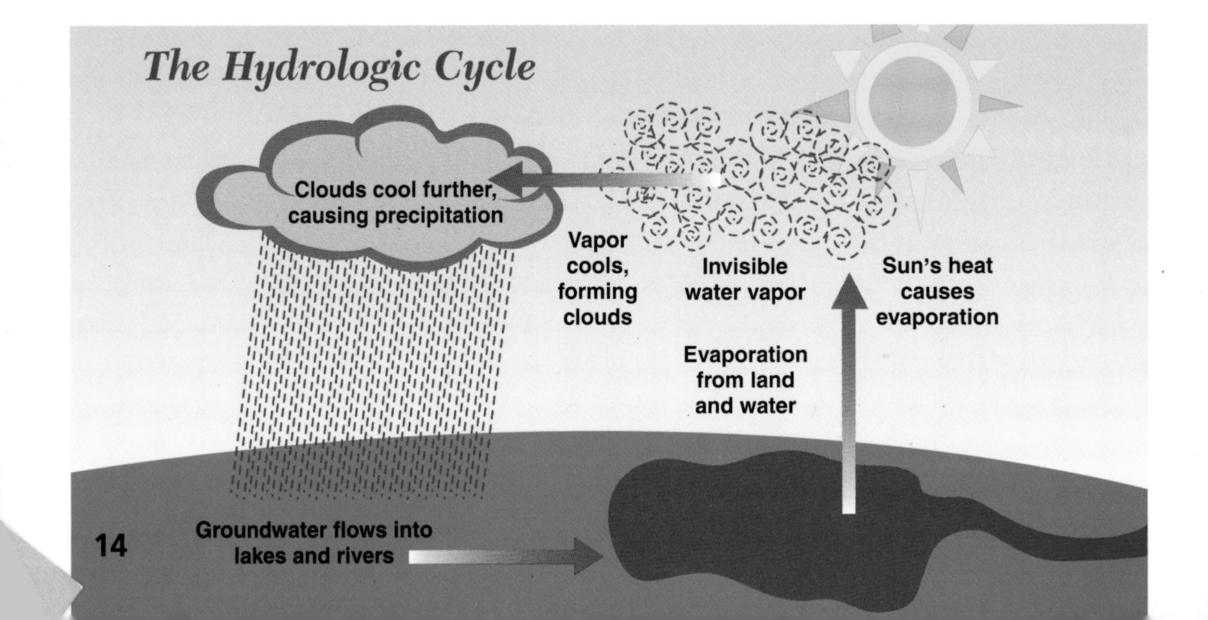

4

Boomers and Twisters

Thunderstorms

Have you ever been startled by a sudden thunderclap? Or maybe you had a split second warning from a sizzling lightning flash just before the boom? Thunderstorms can boil up very quickly when differing weather conditions **collide**.

Thunderstorms happen in many parts of the world. Some are fairly mild. Others are severe. Strong thunderstorms can be dangerous. Lightning can cause burns, destroy power lines, and start fires. Sudden and heavy rain can cause flash flooding.

A Lot of Rain!

When an area gets a lot of rain in a short period of time, a flash flood occurs. These floods usually appear suddenly and leave behind raging waters that rush across the land. Water from flash floods moves very quickly, causing large amounts of damage.

Thunder and lightning can be awesome to observe. The lightning you see is the source of the thunder you hear. When you see lightning but don't hear thunder, it's probably because the storm is too far away.

Lightning is the electricity created when water and ice **particles** collide in a cloud. The lightning flash is very hot—about 50,000°F! The heat instantly warms the air, making it **expand**. The rapidly expanding air sends out a **shock wave** that you hear as thunder.

Whew! How close was that lightning? Count the seconds between the flash and the boom. The lightning is about a mile away for every five seconds in between. If there's not time to count, the lightning is very close. If you can't hear the thunder, it's most likely over ten miles away. Lightning can be seen farther away than thunder can be heard.

Hail

Thunderstorms can also pound the ground with hail. Usually these chunks of ice are pea-sized, but sometimes they're as large as grapefruits. As cold and warm air swirls up and down in a cloud, ice spins around. More and more moisture freezes to the ice pellets until they become too heavy to stay in the air. The hail then falls to the ground, slamming into objects on its way.

Tornadoes

Very strong thunderstorms can produce tornadoes. A tornado is a powerful, swirling, funnel-shaped cloud that sweeps across the land. It can cause great destruction in its path.

Tornadoes can happen almost anytime and anywhere. However, more tornadoes occur in the United States than anyplace else.

Hail

It takes the right combination of ingredients to produce a tornado. Warm air moving over an area with a lot of moisture is the first factor. The warm air is loaded with water and is usually moving away from the equator or between climatic zones. When this air collides with colder air, a tornado forms.

Can you guess the direction the colder air is coming from? It usually arrives from the northern polar region. Sometimes, cold air over a mountain range collides with warmer air to produce a tornado. In the United States, most tornadoes spread from the states along the Gulf of Mexico northward into the Great Plains. The warm air from near the equator picks up moisture as it moves over the Gulf of Mexico. Cold air from the Arctic or the Rocky Mountains meets the humid gulf air, creating tornadoes.

Most of the tornadoes in the United States occur in the spring, usually during the month of May. This is a time when warm air coming from near the equator constantly meets cold air from the Poles and mountains. The air from the North Pole is colder at this time because winter is just ending there. This means there is a larger difference between the two air temperatures during the spring months. The conditions for a twister are just right!

Twisters!

When warm and cold air currents collide, they swirl as they rub or twist together. This motion is where tornadoes get their nickname "twister."

Create your own twister. Fill a plastic 2-liter bottle half full with water. Add food coloring for a more dramatic effect. Cut a strip of duct tape about 3 inches long. Punch a hole in the tape with a paper punch or pencil. Smooth out the tape around the hole's edges so it lies flat. Cover the mouth of the bottle with the piece of duct tape so the hole is over the center of the opening.

Tape another empty 2-liter bottle to the first one. Wrap strips of duct tape tightly around the necks of the bottles. Grab the bottles by the taped necks, and flip them quickly so the one with water is on top. Set the bottles on a table with the empty one on the bottom. Quickly swirl the top bottle in circles parallel to the table. Let go and watch the top half of the bottle to observe your tornado.

5

Peaches, Beaches, and Heat Waves

As spring turns into summer, the warmer weather means a change of activities. Strawberries and peaches sweeten meals. Lawns need mowing. Gardens need watering. Sunscreen, bare feet, and barbecues define the season.

The air and land begin to heat up before summer actually starts. After the vernal equinox, the increase in sunlight slowly begins to warm the hemisphere tilted toward the Sun.

The point at which the axis is closest to the Sun is the year's longest day. This marks the start of the new summer season. This summer **solstice** is the first day of summer. In the Northern Hemisphere, the summer solstice occurs between June 21 and 22. Below the equator, the summer solstice falls between December 22 and 23. If you've lived in the Northern Hemisphere all your life, it might seem strange to know that our friends below the equator think of New Years as a summer holiday!

The warmest days of the season usually occur after the summer solstice. Remember, seasons change gradually. The land and air heat slowly. The amount of heat the hemisphere absorbs from the Sun grows each day through the spring and early summer. Therefore, summer's hottest days are usually after the solstice, even though the hemisphere begins to lean farther away from the Sun. Built-up heat and the daily warmth from the summer Sun combine to make the hottest days in July or August. Long spells of especially hot weather are often called *heat waves*.

A Heat Wave in March or October?

The right conditions can create a really hot day or heat wave anytime from midspring to midfall. Sometimes pockets of air, or **air masses**, can become trapped in an area of high **atmospheric pressure**. With no fresh air flowing in, the trapped air is heated repeatedly, creating a heat wave.

6 Dog Days

Summer isn't all about sunshine, the outdoors, and fresh fruit. In many places around the globe, the summer months are very hot, wet, and stormy.

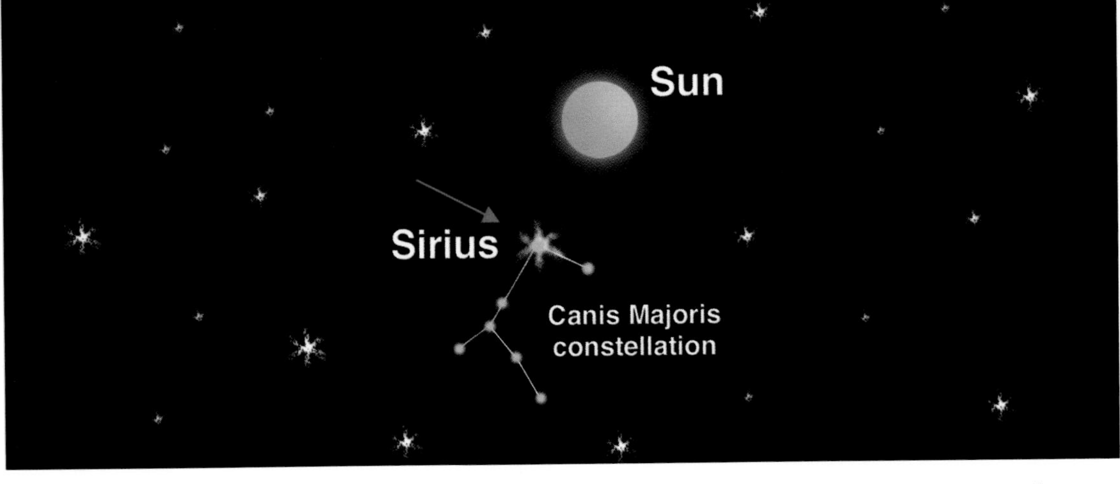

The Dog Days of Summer

Have you ever watched a dog pant and drool on a hot summer day? Did you feel like doing that too? In the Northern Hemisphere, heat waves in late summer are called "dog days." Why is this?

Far out in space, there is a star that is bigger and hotter than our own Sun. Luckily, it is also much farther away. This star is called Sirius. **Astronomers** also call it the Dog Star.

During the late summer in the Northern Hemisphere, the Dog Star rises along with the Sun. Ancient scientists thought Sirius added its heat to the Sun's. They called those especially hot days "the dog days" because of the Dog Star's heat.

The Rainy Season

While areas in the Northern temperate zone are experiencing the dog days, it is a much different story elsewhere. In parts of the

Try This!

Pour two inches of water into the bottom of a glass jar with a lid. Turn the lid upside down and place it on top of the jar. Put several ice cubes in the lid. Observe the changes.

After ten minutes or so, water drops should start forming on the underside of the lid. Why? Some of the water from the bottom of the jar is evaporating and rising to the top of the jar. When the water vapor hits the chilled lid, it condenses, or changes back into a liquid.

tropical zones, summer is the wet season. As the air warms in the tropical zone near the equator, more and more water evaporates and hangs in the air.

Monsoons

Monsoons are winds that bring moist, warm, heavy rainstorms. They occur in areas across India and other parts of Asia throughout the summer months. The monsoons are caused by the tropical wet air from the Indian Ocean being pushed over land by winds passing over the equator. The wet air is forced up against large mountain ranges stretching across the continent, leaving the moisture with nowhere to go but down.

Although the heavy rains and wind can cause flooding and terrible destruction, they are welcomed by Asia's people.

Indian children play in monsoon flooding.

The farmers depend on the water to grow crops. Ninety percent of the water supply in India comes from monsoons. These storms occur from June to September every year.

Seasonal Storms

The word *monsoon* comes from the Arabic word *mawsin*, which means "season." The storms earned this name because they appear every year during the same season.

Tropical Storms

Have you noticed that many extreme weather conditions result from the combination of wind and moisture? and that the equator usually seems to play a part in it? This is also the case with one of nature's most awesome spectacles—hurricanes.

A Storm with Many Names

Hurricanes have different names in different parts of the world. When they form near the Atlantic Ocean, the storms are called *hurricanes*. Over the Indian Ocean, they're called *cyclones*, while in the Pacific Ocean, they're known as *typhoons*. In Australia, many people call them *willie willies*.

Hurricanes are large tropical storms that have **condensed** into smaller storms. But smaller doesn't mean less powerful. The same energy in the large tropical storm is still present, but much more concentrated. Imagine a storm 200 miles wide with 30-mile-per-hour (mph) winds. Suddenly, it squeezes down to just 50 miles wide. Wind speeds jump to over 100 mph or more in just minutes. Now you have a hurricane.

Why do most tropical storms happen in the late summer? Ocean temperatures are at their warmest of the year then. The warm air is heavy with moisture. This air is pushed by winds blowing across the equator from the opposite hemisphere. The warm moisture and winds suck together tightly as the storm travels over the ocean. The storm continues to pick up strength as it moves toward the **shore**, pushing huge waves.

As a hurricane nears the shore, people move **inland** as **coastal** areas are threatened by high seas, wind, and rain. When hurricanes slam full-force into coastal towns, they may cause great damage. The good news is that once on land, the storms die quickly. They also lose strength as they move out of the **tropics** and over cooler ocean water in the temperate zones.

7

Reap What You Sow

As the dog days of summer come to an end, green leaves turn orange, red, and gold. Long days grow shorter. Soon you'll be pulling an extra blanket up around you at night. The shadow that goes everywhere with you stretches longer as the Sun seems to move lower in the sky. (The Sun *is* lower in the sky, but remember, it is the Earth, *not* the Sun, that is moving!). Better gobble up those fresh peaches because they won't be around much longer. The **harvest** is nearly complete.

Canada geese migrate to warmer climates in V-shaped formations.

Some animals frantically store food for the coming cold. Others prepare for long winter naps. A few will move north or south toward a warmer climate as the season changes yet again.

A Long Flight

The arctic tern actually flies from the Arctic to Antarctica and back again during winter to avoid being at the Poles during this freezing season!

Much of nature, recently thriving, now **withers** away. Raking leaves that fall from the skeletons of trees joins the list of yard work. In fact, the falling leaves provide the autumn season with its second name—fall.

The Autumnal Equinox

The autumnal equinox arrives around September 22 in the Northern Hemisphere and March 21 in the Southern Hemisphere. As the Earth continues to move around the Sun, equal days and nights occur again.

The atmosphere gradually cools as winter approaches. But don't put away those shorts just yet. A hot day can hit in fall just when you thought the warm days were over for good! All you need is high pressure to trap a warm air mass, and it will once again feel like summer.

Harvest Weather

Weather can play a large role in the success of a harvest. Drought can deny plants needed moisture and wither developing crops on the vine. Flooding creates the opposite problem. Seeds and crops planted in flooded fields will drown.

Hailstorms hurl ice pellets called *hailstones*. They can beat fruits, vegetables, and surrounding plants down to the ground.

Late rain can be too much of a good thing a little too late. Some fruits and vegetables can rot if late rain leaves them sitting in puddles before they can be harvested. Muddy fields can slow down or stop equipment needed for harvesting and make things even worse.

Of course, the right amount of rain, gentle storms, and dry harvest conditions will produce a bountiful harvest.

8
Indian Summer

With the passing of the autumnal equinox and the arrival of fall, the hemisphere has moved into the coolest half of the year. For the next six months, the hemisphere will be leaning away from the Sun. But cooling in the fall, just as warming in spring, is gradual. Sometimes, well after autumn's arrival, the weather can warm up and stay that way for several days. These warm spells are called Indian summer in some parts of the world and "all-hallown" or "old wives" summer in others.

Indian summer tends to occur in late October or early November in the Northern Hemisphere. The nights may be cool and even bring frost. But the days are much warmer and drier than usual for so late in the season. The name "Indian summer" is believed to have come from the Native Americans' practice of preparing for the coming winter during this time.

Indian summer can occur several times one year and not at all the next. As with other types of weather, it takes the right combination of ingredients to produce an Indian summer. It usually happens when cool polar air bumps against a high pressure air mass and stops moving. As it sits still, it is heated a little more each day. The air may also get hazy and dusty because of the lack of wind.

With no wind, there are no clouds and no flow of cooler air. When this happens, scientists say the air flow has **stagnated**. Under normal conditions, clouds act like a sunshade during the day and like a blanket at night. But a lack of clouds means no shade to keep the **stagnant** air from heating during the day. This creates Indian summer temperatures.

Cooling Off

As the hemisphere continues to move away from the Sun and toward winter, morning frost and overnight freezes become more frequent. The atmosphere cools and wind patterns change, increasing the air flow.

In many places, the rains begin during this cooldown. Away from the tropical zones, rain showers can become snow showers. We're moving toward mittens, sleds, and the shortest day of the year.

9

Long Nights, Snowballs, and Mittens

Long nights. Bare trees. Heavy jackets. Snowmen. Hot chocolate. Plants have gone **dormant** or died back. Bears have holed up and gone to sleep. Most bugs have disappeared (except for the ones that sneak into the house to keep warm!). Winter has arrived.

The Winter Solstice

We have reached the winter solstice. The Earth's tilt and orbit have brought the hemisphere to the farthest point from the Sun. It happens between December 21 and 22 above the equator and June 21 and 22 below. The farther away from the equator, the colder the temperatures become.

Just as the summer solstice is the longest day of the year, the winter solstice is the shortest. In fact, in the polar regions, there are stretches of time near each solstice when the Sun never fully rises or sets.

Thousands of years ago, before people started studying the cycles of the Earth and Sun, the winter solstice was a scary time. The days kept getting shorter. No one knew how long the darkness would last. Would the Sun simply disappear one day, leaving the Earth in total darkness?

But after the solstice, the days would begin to lengthen again, and everything grew warmer. Because of this, many winter holidays, including Christmas, began as solstice celebrations. The winter solstice marked the beginning of a new and hopeful year.

Try This!

You will need a flashlight, a dark piece of paper, and a tilting globe.

In a darkened room, hold the flashlight about six inches above the paper and point it straight down. What shape does the light form? Gradually tilt the flashlight. How does the shape of the light change? Is it brighter or dimmer?

Now aim the flashlight at the globe. Point straight at the equator. Move the globe so that the axis, or tilt, is leaning away from the flashlight. Notice how one Pole is getting light all the time while the other Pole is getting very little?

The light on the paper shows how the Sun's light and heat are stronger and weaker as the angle of sunlight changes with the seasons. The light on the globe shows how polar regions experience days of 24-hour light and days of 24-hour darkness.

10

A Time of Water

The name *winter* comes from an old European word that means "time of water." Without winter's rainstorms, blizzards, and snowcapped peaks, the faucets and fountains would run dry.

The Water Cycle

The hydrologic cycle guides all weather. During the winter, colder air flows outward from the Poles. The atmosphere gradually cools until wind patterns change. Winds pick up moisture as they travel, creating the storms that will replace water on the Earth's surface. Rain refills underground pockets of water, lakes, and rivers. Snow that falls and collects during the winter will melt to feed springs, streams, and rivers throughout the summer.

Water never disappears. It just changes form and is moved around the world by weather. When water evaporates, it rises into the air as a gas, or vapor. Eventually, it joins with more water vapor to form a cloud. The cloud is carried on wind **currents** and can be heated or cooled. It can also gather more water until wind, elevation, and temperature are right for the water to fall as rain or snow.

As the atmosphere continues to cool, air masses change direction and height. Air cools more rapidly than water does. As colder air mixes with warmer air and flows over bodies of warm water, the warm water adds moisture while the colder air holds less moisture. Stormy weather relieves the air of its extra moisture.

Ocean currents near the equator send moisture to the tropics. It mixes with the colder air flowing from the Poles. This causes precipitation in coastal areas. The moisture travels inward toward mountain ranges. Rain falls along the coasts, while snow drops in higher mountain elevations. Often, most of the rain or snow falls on the side of the mountain facing the coast. The far side might only receive cold air and wind. This is because the cold

air at the top of the mountain causes the warm, moist air to condense and fall as rain before passing over the mountain to the other side.

Back to the Equator

The equator is the engine that drives our weather machine. It constantly churns out heat and moisture, which move around the world setting weather patterns in motion. Scientists have found superstrong bursts of warm wind and moisture that start in the tropical regions and travel around the globe.

These weather-making bursts occur most often from December through May. It takes from one to three months for these bursts to journey around the world. As they travel, they can trigger everything from summer monsoons to winter blizzards.

El Niño

Every few years, these bursts overheat Pacific Ocean waters near the equator. Strange weather patterns result. Scientists call this warming El Niño, which means "Christ child." This name was chosen because the unusually warm ocean currents tend to occur at Christmastime in the Northern Hemisphere. El Niño is most noticeable when it happens during December or January.

El Niño blows in, damaging an oceanfront restaurant.

El Niño seems to cause unusually stormy weather in places like Chile, on the southwest side of South America. In other areas, including Brazil and Australia, El Niño is believed to cause drought. Researchers aren't sure how El Niño affects the weather in other places. Some think it makes the western United States wetter than usual in the winter. Other scientists believe just the opposite.

Why is El Niño so hard to figure out? Time and distance create problems for scientists. The unusual weather happens six months to a year after the warm ocean currents are detected. And there is a great distance between where El Niño begins and where the unusual weather occurs. With the passage of so much time and distance, many things could have happened to change the weather. Scientists continue to study this unusual weather system to better understand and predict its effects.

Accurate weather records have only been kept for about 100 years. Scientists know they need more information to better predict the weather and learn its patterns. But while they continue to study and learn, the Earth will keep spinning and the seasons will keep changing.

Internet Connections and Related Reading for Seasons and Patterns

http://familyeducation.com/topic/front/0,1156,1-4205,00.html
Explore the winter solstice through quizzes, activities, and information about ancient myths.

http://www.infoplease.com/spot/riteofspring1.html
Learn more about the equinoxes and the seasons at this site.

http://www.allaboutnature.com/subjects/astronomy/planets/earth/seasons.shtml
Check out the diagram of the seasons and the axis's tilt here. Read more about the solstices and equinoxes, and take a quiz about the seasons.

http://www.scienceu.com/observatory/articles/seasons/seasons.html
The "reasons for the seasons" will be clear to you after you see the animated diagrams and read the information at this site.

http://www.howstuffworks.com/category-weather.htm
Discover how all that weather "stuff" works at this science site. You can dive into the seasons, weather, hurricanes, lightning, floods, tornadoes, and more.

❊❊❊❊❊❊❊❊❊❊❊❊❊❊❊❊❊❊❊❊❊❊❊❊❊❊❊❊❊❊

Down Comes the Rain by Franklyn M. Branley. Answers the question "Where does the rain come from?" and suggests activities that will reveal the water cycle in action. HarperCollins 1997. [RL 2 IL K–3] (5540701 PB 5540702 CC)

Sunshine Makes the Seasons by Franklyn M. Branley. Describes how sunshine and the tilt of the Earth's axis are responsible for the changing seasons. HarperCollins, 1985. [RL 2 IL K–3] (8746401 PB 8746402 CC)

Thirteen Moons on Turtle's Back: A Native American Year of Moons by Joseph Bruchac and Jonathan London. In many Native American cultures it is believed that the 13 scales on Turtle's shell stand for the 13 cycles of the Moon, each with its own name and a story that relates to the changing seasons. Putnam, 1997. [RL 3 IL K–6] (5549801 PB 5549802 CC)

Weather by Seymour Simon. Explores the causes, changing patterns, and forecasting of weather. HarperCollins, 2000. [RL 6.3 IL K–5] (3182601 PB 3182602 CC)

Weather: Poems for All Seasons selected by Lee Bennett Hopkins. Poems for rain, sleet, snow, or Sun by such poets as Langston Hughes, Ogden Nash, Valerie Worth, and others. HarperCollins, 1995. [RL 2 IL 1–4] (4799301 PB 4799302 CC)

- RL = Reading Level
- IL = Interest Level

Perfection Learning's catalog numbers are included for your ordering convenience. PB indicates paperback. CC indicates Cover Craft. HB indicates hardback.

Glossary

absorb (uhb SORB) take in

air mass (air mas) huge body of air that has similar characteristics throughout

astronomer (uh STRAH nuh mer) person who studies the stars and planets

atmosphere (AT muhs fear) air surrounding the Earth

atmospheric pressure (at muhs FEAR ik PRESH er) weight of the air on the Earth's surface

axis (AK sis) imaginary line running through the Earth

barren (BAIR uhn) not able to grow living things

climate (KLEYE mit) weather

climatic (kleye MAT ik) having to do with climates

coastal (KOHS tuhl) having to do with the land near a coast, or shore (see separate entry for *shore*)

collide (kuh LEYED) come together with great force; bump or smash into

condensed (kuhn DENSD) made dense or more compact; pushed together

current (KER ent) stream of air or water moving continuously in one direction

dormant (DOR muhnt) experiencing a lack of movement or growth

elevation (el uh VAY shuhn) height

environment (en VEYE er muhnt) set of conditions found in a certain area, such as type of land, weather, plants, etc.

evaporate (ee VAP or ayt) to change from a liquid to a gas

expand (ik SPAND) to grow larger

extreme (ik STREEM) going beyond the usual, ordinary, or expected

harvest (HAR vest) the process of gathering a crop

hemisphere (HEM uh sfear) half of the Earth

humid (HYOU mid) moist; wet

hydrologic cycle (heye druh LAH jik SEYE kuhl) water cycle on Earth

inland (IN luhnd) away from the shore (see separate entry for *shore*)

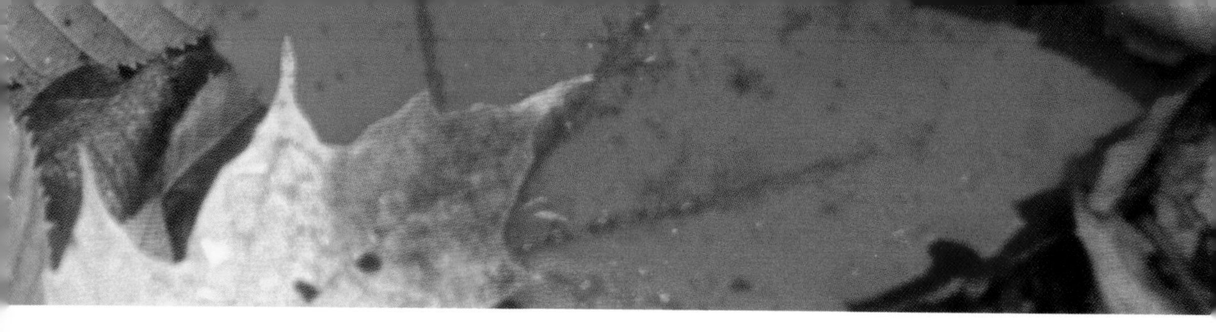

orbit (OR bit) the path of an object as it revolves, or moves in a circular path, around another object

particle (PAR tuh kuhl) small piece of something

polar (POH ler) having to do with the North and/or South Poles

precipitation (pree sip i TAY shuhn) moisture that falls to the ground (rain, snow, hail, sleet, etc.)

reflect (ree FLEKT) bounce off

shock wave (shahk wayv) wave of sound caused by a collision

shore (shor) land bordering, or next to, a body of water

solar system (SOH ler SIS tuhm) the Sun or another star and the celestial bodies that revolve around it

solstice (SOHL stis) two points during the Earth's orbit where the equator is the greatest distance from the Sun

sphere (sfear) ball; globe

stagnant (STAG nuhnt) not flowing, moving, or developing

stagnated (STAG nay ted) stopped flowing, moving, or developing

tropics (TROP iks) regions right above and below the equator

wither (WITH er) to shrivel and dry up

Index